JIBUNOH'S METHOD FOR EVALUATNG THE DETERMINANT OF AN N×N MATRIX

A monograph on research discovery

C.C. JIBUNOH
B.Sc, M.Ed, M.Sc, D.Sc, MNMS, MNSA.

All Rights Reserved
*No part of this publication may be reproduced, by any means,
Be it electronic, mechanical or otherwise, without the
Prior permission in writing from the copyright owner*

ISBN: 978-1494290931

First published in 2009

© C. C. JIBUNOH

Email: chafachid@gmail.com GSM: +2347063860446, +2348053336496
Department of Mathematics and Statistics
Delta State Polytechnic, Ogwashi-Uku
Nigeria

DEDICATION

TO MY SONS CHIKWADOM AND CHICHEBEM
WHO ARE YOUNGER AND ZEALOUS MATHEMATICIANS

PREFACE

It has been a normal practice to publish a research discovery in an appropriate Journal meant for specialists in the field of the research or other persons who may be interested in the results. In many cases, not every specialist gets in touch with the Journal, especially if it is not in regular circulation.

Where a research result is not mainly of specialists' interest but addresses a problem that cuts across many disciplines, I am of the opinion that the discovery could also be published as a monograph which can be circulated promptly by a publisher. The paper 'The Determinant of an n x n matrix by reduction to echelon form' is a baby of many disciplines and has been in high demand since it was presented to the Nigerian Mathematical Society (NMS), in the Annual Conference of June 2009, at Ilorin, Nigeria. Rather than responding to the requests for photocopies of the paper, I considered it appropriate to place the exact paper in a publication as a monograph. As pointed out, the idea of a determinant of a matrix is not mainly for mathematicians or statisticians *per se*. It is useful to engineers, technologists, scientists, social scientists, computer scientists, business administrators, accountants or other persons involved in matrices, quantitative techniques or empirical evaluations. The knowledge of determinant is required at all levels of study, from school certificate level to post graduate studies and beyond.

The spectacular aspect of the monograph is that it presents a new, simpler and systematic method of evaluating the determinant of an n x n matrix, no matter how large the value of n. The traditional method which uses minor determinants or cofactors is often tedious when n is large and is now outdated. This new and simple method which is a product of research may be referred to as <u>*Jibunoh's method for evaluating the determinant of an n x n matrix.*</u>

I wish to thank many senior colleagues of the Nigerian Mathematical Society for encouraging me to reproduce this paper for wider dissemination.

Dr. C.C. Jibunoh
Ogwashi-Uku, Nigeria
15th July, 2009

ARRANGEMENT OF SECTIONS

	Page
Abstract	1
1. Introduction	1
2. Formulations for the 2 x 2 and 3 x 3 general matrices and extension to the general n x n matrices	2
3. Examples	8
4. Conclusion	21
5. Exercises	21
References	24
Index	25

DETERMINANT OF AN N × N MATRIX BY REDUCTION TO ECHELON FORM

BY

Chafa C. Jibunoh
Email: chafa_chid@yahoo.com GSM: +2348053336496
**Department of Mathematics and Statistics
Delta State Polytechnic Ogwashi-Uku
Nigeria**

As delivered to the Nigerian Mathematical Society (NMS) in the conference of June 2009, at Ilorin, Nigeria.

Abstract

A simple and systematic procedure for obtaining the determinant of any n x n matrix without the use of cofactors is developed in this paper. The procedure involves reduction of the square matrix to echelon form and dividing the product of the diagonal elements by the product of some numbers defined as lower multipliers. The simplicity of the procedure is maintained and appreciated as the order of the matrix, $n \to \infty$. This makes it superior to the traditional method of cofactors which is often tedious to apply in high order matrices.

1. Introduction

Multiplying the element of any row or column by their respective cofactors and summing the products, constitute the traditional method of obtaining the determinant of a square matrix. When the order n of the matrix is more than 3 this traditional method becomes cumbersome. The rule of Sarrus is restricted only to the case n = 3. However several rules based on elementary row or column operations abound in the literature. These tend to reduce the order of the matrix from n to n-1, but none of these has a unique formulation for deriving the determinant of a general n x n matrix.

In this paper, we shall formulate a systematic and unique rule for obtaining the determinant of any n x n matrix via a simple reduction of the matrix to echelon form. This rule is general, simple and efficient as $n \to \infty$.

We shall derive the formulation for 2 x 2 and 3 x 3 general matrices and prove that the rule is general for n x n matrices.

2. Formulations for the 2 x 2 and 3 x 3 general matrices and extension to the general n x n matrices.

Let a 2 x 2 matrix be given by

$$A = \begin{pmatrix} a_{11} & a_{12} \\ a_{21} & a_{22} \end{pmatrix} \quad (2.1)$$

where a_{ij}, are real or complex entries.

Then obviously the determinant

$$\det A = a_{11}a_{22} - a_{21}a_{12} \quad (2.2)$$

Now multiply Row 1 by a_{21} and Row 2 by a_{11}, which procedure is indicated by the scheme.

$$\begin{pmatrix} a_{11} & a_{12} \\ a_{21} & a_{22} \end{pmatrix} \qquad \begin{matrix} a_{21} \\ (a_{11}) \end{matrix} \quad (2.3)$$

The elementary row operation a_{11} Row 2 – a_{21} Row 1, converts (2.3) to

$$\begin{pmatrix} a_{11} & a_{12} \\ 0 & a_{11}a_{22} - a_{21}a_{12} \end{pmatrix} \quad (2.4)$$

We define (a_{11}) as the lower multiplier for the operation

The matrix (2.4) is the matrix **A** reduced to echelon form. The determinant of (2.4) is given by the product of the diagonal elements i.e.

$$a_{11}(a_{11}a_{22} - a_{21}a_{12}) \quad (2.5)$$

Comparing (2.5) with (2.2) we find that det **A** is obtained by dividing (2.5) by the lower multiplier (a_{11}), i.e.

$$\det A = \frac{a_{11}(a_{11}a_{22} - a_{21}a_{12})}{a_{11}} \quad (2.6)$$

$$= a_{11}a_{22} - a_{21}a_{12}$$

Consider a general 3 x 3 matrix

$$B = \begin{pmatrix} a_{11} & a_{12} & a_{13} \\ a_{21} & a_{22} & a_{23} \\ a_{31} & a_{32} & a_{33} \end{pmatrix}$$

Applying the chess board rule of signs and expanding with the first row in the traditional way, we have

$$\det \mathbf{B} = a_{11} a_{22} a_{33} - a_{11} a_{32} a_{23} - a_{12} a_{21} a_{33}$$
$$+ a_{12} a_{31} a_{23} + a_{13} a_{21} a_{32} - a_{13} a_{31} a_{22} \tag{2.7}$$

To reduce **B** to echelon form, a similar procedure is adopted as in the 2 x 2 matrix i.e.

a_{11}	a_{12}	a_{13}	a_{21}	a_{31}	
a_{21}	a_{22}	a_{23}	(a_{11})		
a_{31}	a_{32}	a_{33}		(a_{11})	

(2.8)

The multipliers required to convert the positions of a_{21} and a_{31} to zeros are indicated on the RHS of the matrix in two separate columns, where for the first operation (a_{11}) is the lower multiplier and for the second operation (a_{11}) also is the lower multiplier. The first row is defined as the pivot row and remains unaltered till the end of the operations. The elementary row operations are now

a_{11} Row 2 – a_{21} Row 1

a_{11} Row 3 – a_{31} Row 1

Hence the reduced matrix becomes

a_{11}	a_{12}	a_{13}	
0	$a_{11}a_{22} - a_{12}a_{21}$	$a_{11}a_{23} - a_{13}a_{21}$	$a_{11}a_{32} - a_{12}a_{31}$
0	$a_{11}a_{32} - a_{12}a_{31}$	$a_{11}a_{33} - a_{13}a_{31}$	$(a_{11}a_{22} - a_{12}a_{21})$

(2.9)

On the RHS of the reduced matrix above, are the multipliers for the next reduction using the new second row as the pivot row. The lower multiplier is $(a_{11}a_{22} - a_{12}a_{21})$. Hence the matrix **B** reduced to echelon form becomes

$$C = \begin{pmatrix} a_{11} & a_{12} & a_{13} \\ 0 & a_{11}a_{22}-a_{12}a_{21} & a_{11}a_{23} - a_{13}a_{21} \\ 0 & 0 & [(a_{11}a_{33} - a_{13}a_{31})(a_{11}a_{22} - a_{12}a_{21}) \\ & & -(a_{11}a_{23} - a_{13}a_{21})(a_{11}a_{32} - a_{12}a_{31})] \end{pmatrix} \quad (2.10)$$

The determinant of **C**, the matrix in echelon form is the product of the diagonal elements and is given after simplification by

$a_{11}^2 (a_{11}a_{22} - a_{12}a_{21})(a_{11}a_{22}a_{33} - a_{11}a_{32}a_{23} - a_{12}a_{21}a_{33} + a_{12}a_{31}a_{23} + a_{13}a_{21}a_{32} - a_{13}a_{31}a_{22})$

$= a_{11}^2 (a_{11}a_{22} - a_{12}a_{21}) \det \mathbf{B}$ \hfill (2.11)

The product of the lower multipliers from (2.8) and (2.9) is

$a_{11}^2 (a_{11}a_{22} - a_{12}a_{21})$ \hfill (2.12)

Hence det **B** is obtained by dividing (2.11) by the product of lower multipliers. We shall establish a general rule for all square matrices of order n.

Let A_n be a general square matrix of order n denoted by

$$A_n = \begin{pmatrix} a_{11} & a_{12} & \cdots & a_{1n} \\ a_{21} & a_{22} & \cdots & a_{2n} \\ \cdots & \cdots & \cdots & \cdots \\ \cdots & \cdots & \cdots & \cdots \\ a_{n1} & a_{n2} & \cdots & a_{nn} \end{pmatrix} \quad (2.13)$$

From the matrix (2.13), we may extract the square matrices; $A_2, A_3, \ldots, A_k$, $k \leq n$

Let us reformulate the reductions of the matrices of different orders as follows;

Order 2

a_{11}	a_{12}	a_{21}
a_{21}	a_{22}	(a_{11})
a_{11}	a_{12}	
0	b_{22}	$= \begin{pmatrix} a_{11} & a_{12} \\ 0 & e_{22} \end{pmatrix}$

(2.14)

By backward substitution

$$e_{22} = b_{22} = a_{11} a_{22} - a_{12} a_{21} = \det A_2$$

Thus $\det A_2 = \dfrac{a_{11} e_{22}}{a_{11}}$, as in (2.6)

Order 3

a_{11}	a_{12}	a_{13}	a_{21}	a_{31}
a_{21}	a_{22}	a_{23}	(a_{11})	
a_{31}	a_{32}	a_{33}		(a_{11})
a_{11}	a_{12}	a_{13}		
0	$\det A_2$	b_{23}	b_{32}	
0	b_{32}	b_{33}	$(\det A_2)$	

5

$$\begin{vmatrix} a_{11} & a_{12} & a_{13} \\ 0 & \det A_2 & b_{23} \\ 0 & 0 & c_{33} \end{vmatrix} = \begin{pmatrix} a_{11} & a_{12} & a_{13} \\ 0 & \det A_2 & b_{23} \\ 0 & 0 & e_{33} \end{pmatrix} \qquad (2.15)$$

Then by backward substitutions

$e_{33} = c_{33} = b_{33} \det A_2 - b_{23} b_{32}$

$= (a_{33} a_{11} - a_{13} a_{31}) \det A_2 - [(a_{23} a_{11} - a_{13} a_{21})(a_{32} a_{11} - a_{12} a_{31})]$

$= a_{11} \det A_3 \qquad (2.16)$

Hence $\det A_3 = \dfrac{a_{11} \det A_2 e_{33}}{a_{11}^{2} \det A_2}$

where $a_{11} \det A_2 \, e_{33}$ is the product of the diagonal elements of the echelon matrix (2.15) and $a_{11}^2 \det A_2$ is the product of the lower multipliers of the reductions.

Order 4

a_{11}	a_{12}	a_{13}	a_{14}	a_{21}	a_{31}	a_{41}
a_{21}	a_{22}	a_{23}	a_{24}	(a_{11})		
a_{31}	a_{32}	a_{33}	a_{34}		(a_{11})	
a_{41}	a_{42}	a_{43}	a_{44}			(a_{11})

a_{11}	a_{12}	a_{13}	a_{14}			
0	$\det A_2$	b_{23}	b_{24}	b_{32}	b_{42}	
0	b_{32}	b_{33}	b_{34}	$(\det A_2)$		
0	b_{42}	b_{43}	b_{44}		$(\det A_2)$	

a_{11}	a_{12}	a_{13}	a_{14}		
0	$\det A_2$	b_{23}	b_{24}		
0	0	c_{33}	c_{34}	c_{43}	
0	0	c_{43}	c_{44}	(c_{33})	

$$\begin{array}{|cccc|} \hline a_{11} & a_{12} & a_{13} & a_{14} \\ 0 & \det A_2 & b_{23} & b_{24} \\ 0 & 0 & c_{33} & c_{34} \\ 0 & 0 & 0 & d_{44} \\ \hline \end{array} = \begin{pmatrix} a_{11} & a_{12} & a_{13} & a_{14} \\ 0 & \det A_2 & b_{23} & b_{24} \\ 0 & 0 & e_{34} & c_{34} \\ 0 & 0 & 0 & e_{44} \end{pmatrix} \quad (2.17)$$

From above, we have

$c_{33} = e_{34}$ and $d_{44} = e_{44}$

By backward substitution, we find

$$e_{44} = a_{11}^2 \det A_2 \det A_4 \qquad (2.18)$$

Hence,

$$\det A_4 = \frac{a_{11} \det A_2 e_{34} e_{44}}{a_{11}^3 (\det A_2)^2 e_{34}} \qquad (2.19)$$

A similar reduction for order 5, gives

$$\det A_5 = \frac{a_{11} \det A_2 e_{35} e_{45} e_{55}}{a_{11}^4 (\det A_2)^3 (e_{35})^2 e_{45}} \qquad (2.20)$$

where, using backward substitutions, we obtain,

$$e_{55} = a_{11}^3 (\det A_2)^2 e_{35} \det A_5 \qquad (2.21)$$

Now for any n x n matrix it will follow from the above formulations of orders 2 to 5, that the product of lower multipliers shall be given by

$$a_{11}^{n-1} (\det A_2)^{n-2} (e_{3n})^{n-3} (e_{4n})^{n-4} \ldots e_{n-1,n} \qquad (2.22)$$

since a_{11} will appear (n-1) times as lower multipliers, $\det A_2$ will appear (n-2) times, etc. In this case;

$$\det A_n = \frac{a_{11} \det A_2 e_{3n} e_{4n} \ldots e_{n-1,n} e_{nn}}{a_{11}^{n-1} (\det A_2)^{n-2} (e_{3n})^{n-3} (e_{4n})^{n-4} \ldots e_{n-1,n}} \qquad (2.23)$$

where $a_{11} \neq 0$, $\det A_2 \neq 0$ and by assumption $e_{j,n} \neq 0$ for all j. This equation will not be true except by backward substitutions;

$$e_{nn} = \det A_n \, a_{11}^{n-2} (\det A_2)^{n-3} (e_{3n})^{n-4} \ldots 1, \qquad (2.24)$$

for all n. This is the general formula for e_{nn} and is validated for n=2 to 5.

Hence, by reduction to echelon form, $\det A_n$ is given by the general formula (2.23), for all n. The formulas (2.23) and (2.24) assume that all n-1, stages of the reductions are completed and no row is jumped. The formulas also assume that the elementary row operations leading to the matrix in echelon form are carried out directly on the matrix entries and not by factors obtained by using the LCMs of the entries. However LCMs are allowed and often do simplify the reductions to echelon form.

Remarks

In the above proof or deductions, we assumed that $a_{11} \neq 0$ and $\det A_2 \neq 0$. If $a_{11}=0$, the row (or column) containing it can be interchanged with another row or column which will have a non-zero entry for a_{11}. Similarly $\det A_2$ can be made non-zero, also by certain row or column, interchanges. For any square matrix of order n, the reductions to echelon form take $\leq$ n-1, stages, if the practical procedure is used and exactly n-1, stages, if the theoretical procedure is used – see Examples 3, 4 and 5. By assumption, $e_{j,n}$ is non-zero as a lower multiplier, or it can be made non-zero, by interchange of rows or columns, for any $j \leq$ n-1. But for some j, it can appear as a zero entry (or entries) in the numerator of (2.23), among the diagonal elements of the matrix in echelon form, which may cause the appearance of such zero entries in the denominator of (2.23). In this case, by definition, $\det A_n = 0$.

The general formulation for an n x n matrix is therefore summarized as follows; Let **A** be any general n x n matrix. Reduce **A** by elementary row operations as described above, to a matrix in echelon form denoted by **C**, then,

$$\det \mathbf{A} = \frac{\text{product of the diagonal elements of } \mathbf{C}}{\text{product of lower multipliers from each stage of the reduction to } \mathbf{C}} \quad (2.25)$$

3.
Examples

Example 1

$$\text{Let } \mathbf{A} = \begin{pmatrix} 2 & 3 \\ 5 & 6 \end{pmatrix}$$

Then

2	3	5
5	6	
		(2)

leads to

$$C = \begin{pmatrix} 2 & 3 \\ 0 & -3 \end{pmatrix}$$

Hence det $A = \dfrac{-6}{2} = -3$

Example 2

Let $A = \begin{pmatrix} 1 & 2 & 3 \\ 4 & 5 & 6 \\ 7 & 8 & 5 \end{pmatrix}$

Then

1	2	3	4	7
4	5	6	(1)	
7	8	5		(1)
1	2	3		
0	-3	-6		2
0	-6	-16	(1)	

which leads to

$$C = \begin{pmatrix} 1 & 2 & 3 \\ 0 & -3 & -6 \\ 0 & 0 & -4 \end{pmatrix}$$

Hence

$$\det A = \frac{1 \times -3 \times -4}{1 \times 1 \times 1} = 12$$

Observe that in the 2nd stage of reduction of the matrix **A**, the LCM of 3 and 6 was used to obtain the multipliers 2 and 1, with (1) as the lower multiplier. The use of LCM whenever possible simplifies the reduction.

Note: The use of LCMs is part of the practical procedure of reductions to echelon form. Also for practical reductions to echelon form, add the rows when the leading elements have opposite signs, otherwise subtract the upper from the subsequent row (with the application of LCMs, where possible). The theory does not assume the use of LCMs. It requires the use of the exact leading entries (with the signs) as multipliers and relies only on subtractions. The practical and theoretical procedures lead to the same result. But to verify the formulas (2.23) and (2.24) the <u>theoretical</u> procedure of reductions must be used.

Example 3

$$A = \begin{pmatrix} 2 & 0 & -1 \\ 3 & 0 & 2 \\ 4 & -3 & 7 \end{pmatrix}$$

as taken from [2].

Obviously to make $detA_2$ non-zero, as required by (2.23), we can interchange the 2nd and 3rd <u>columns</u> and have the following reductions, by <u>the practical procedure</u>

2	-1	0	3	2
3	2	0	(2)	
4	7	-3		(1)
2	-1	0		
0	7	0	9	
0	9	-3	(7)	

which lead to

$$C = \begin{pmatrix} 2 & -1 & 0 \\ 0 & 7 & 0 \\ 0 & 0 & -21 \end{pmatrix}$$

It is known that the interchange of two columns (or rows) of any arbitrary square matrix **A**, introduces a negative sign to the determinant.

Hence

$$\det \mathbf{A} = \frac{(-1)^1 \times 2 \times 7 \times -21}{2 \times 1 \times 7} = 21$$

The index on (- 1) is 1, since there was only one interchange of columns (or rows) in the reductions. For k interchanges in the process we multiply by $(-1)^k$.

*This example shows that column interchanges are as good as row interchanges, where necessary.

Example 4

Let $\mathbf{A} = \begin{pmatrix} 5 & 4 & 2 & 1 \\ 2 & 3 & 1 & -2 \\ -5 & -7 & -3 & 9 \\ 1 & -2 & -1 & 4 \end{pmatrix}$, a 4x4 matrix, also taken from [2]

(a) By <u>the practical procedure</u> we obtain the following:

5	4	2	1	2	1	1
2	3	1	-2	(5)		
-5	-7	-3	9		(1)	
1	-2	-1	4			(5)
5	4	2	1			
0	7	1	-12	3	2	
0	-3	-1	10	(7)		
0	-14	-7	19		(1)	

5	4	2	1	
0	7	1	-12	
0	0	-4	34	5
0	0	-5	-5	(4)

Hence

$$C = \begin{pmatrix} 5 & 4 & 2 & 1 \\ 0 & 7 & 1 & -12 \\ 0 & 0 & -4 & 34 \\ 0 & 0 & 0 & -190 \end{pmatrix}$$

$$\det A = \frac{5 \times 7 \times -4 \times -190}{5 \times 1 \times 5 \times 7 \times 1 \times 4} = 38$$

(b) Alternatively, for this Example 4, we obtain the following, by <u>the theoretical procedure</u>:

5	4	2	1	2	-5	1	
2	3	1	-2	(5)			
-5	-7	-3	9		(5)		
1	-2	-1	4			(5)	
5	4	2	1				
0	7	1	-12	-15	-14		
0	-15	-5	50	(7)			
0	-14	-7	19		(7)		

5	4	2	1	
0	7	1	-12	
0	0	-20	170	-35
0	0	-35	-35	(-20)

Hence

$$C = \begin{pmatrix} 5 & 4 & 2 & 1 \\ 0 & 7 & 1 & -12 \\ 0 & 0 & -20 & 170 \\ 0 & 0 & 0 & 6650 \end{pmatrix}$$

$$\det A = \frac{5 \times 7 \times -20 \times 6650}{5^3 \times 7^2 \times -20} = 38$$

The formulas (2.23) and (2.24) may now be verified using the diagonal elements of **C**. Manually, the practical procedure is simpler than the theoretical procedure but for computer evaluation of the determinant, the theoretical procedure is preferable, since det **A** can be found automatically using only the diagonal elements of **C**.

Example 5

$$\text{Let} \quad \mathbf{A} = \begin{pmatrix} 2 & 1 & 0 & 0 \\ 0 & 2 & 0 & 0 \\ 0 & 0 & 1 & 1 \\ 0 & 0 & -2 & 4 \end{pmatrix}$$

as taken from [2]

(a) Then by the practical procedure we obtain;

2	1	0	0	
0	2	0	0	
0	0	1	1	2
0	0	-2	4	(1)

Hence

$$\mathbf{C} = \begin{pmatrix} 2 & 1 & 0 & 0 \\ 0 & 2 & 0 & 0 \\ 0 & 0 & 1 & 1 \\ 0 & 0 & 0 & 6 \end{pmatrix}$$

This example shows that (by the practical procedure) the rows of a matrix which are already in the echelon format need no longer be reduced and can be jumped. Hence there is only one reduction using the third row as the pivot row, to obtain the matrix **C**.

Therefore

$$\det \mathbf{A} = \frac{2 \times 2 \times 1 \times 6}{1} = 24$$

The inference is clear that if the original matrix **A** is, for example, already in complete echelon form, then practically, there will be no reduction and the determinant of **A** will simply be the product of the diagonal elements. This agrees with the traditional evaluation.

(b) By the theoretical procedure on this Example 5, we obtain:

2	1	0	0	0	0	0
0	2	0	0	(2)		
0	0	1	1		(2)	
0	0	-2	4			(2)
2	1	0	0			
0	4	0	0	0	0	
0	0	2	2	(4)		
0	0	-4	8		(4)	
2	1	0	0			
0	4	0	0			
0	0	8	8	-16		
0	0	-16	32	(8)		

Hence

$$C = \begin{pmatrix} 2 & 1 & 0 & 0 \\ 0 & 4 & 0 & 0 \\ 0 & 0 & 8 & 8 \\ 0 & 0 & 0 & 384 \end{pmatrix}$$

$$\det A = \frac{2 \times 4 \times 8 \times 384}{2^3 \times 4^2 \times 8} = 24, \text{ as before.}$$

As seen above, the theoretical procedure does not allow the jumping of any row. A striking aspect is that even a matrix A_n, originally in echelon form can also be reduced in n-1, stages, to echelon form, by the <u>theoretical procedure</u> and the formula (2.23), applied to obtain the determinant.

Example 6

$$A = \begin{pmatrix} 0 & 2 & 1 & 4 & -1 & 3 \\ 1 & 2 & -1 & 3 & 4 & 0 \\ 0 & 1 & 1 & -1 & 2 & -1 \\ 2 & 3 & -4 & 2 & 0 & 5 \\ 1 & 1 & 1 & 3 & 0 & 2 \\ -1 & -1 & 2 & -1 & 2 & 0 \end{pmatrix}$$

This is a square matrix of order 6, taken from Burden and Faires [1], p.371, where the determinant of this matrix was found by the traditional method. We interchange the first and second rows to have a number other than zero, to lead the pivot row, since by (2.23), a_{11} must be non-zero. This interchange of two rows, of course, means that the eventual determinant shall be multiplied by (-1). Hence we have the following reductions:

1	2	-1	3	4	0	2	1	1
0	2	1	4	-1	3			
0	1	1	-1	2	-1			
2	3	-4	2	0	5	(1)		
1	1	1	3	0	2		(1)	
-1	-1	2	-1	2	0			(1)

$$\begin{array}{cccccc|cccc}
1 & 2 & -1 & 3 & 4 & 0 & & & & \\
0 & 2 & 1 & 4 & -1 & 3 & 1 & 1 & 1 & 1 \\
0 & 1 & 1 & -1 & 2 & -1 & (2) & & & \\
0 & -1 & -2 & -4 & -8 & 5 & & (2) & & \\
0 & -1 & 2 & 0 & -4 & 2 & & & (2) & \\
0 & 1 & 1 & 2 & 6 & 0 & & & & (2) \\
\hline
1 & 2 & -1 & 3 & 4 & 0 & & & & \\
0 & 2 & 1 & 4 & -1 & 3 & & & & \\
0 & 0 & 1 & -6 & 5 & -5 & 3 & 5 & 1 & \\
0 & 0 & -3 & -4 & -17 & 13 & (1) & & & \\
0 & 0 & 5 & 4 & -9 & 7 & & (1) & & \\
0 & 0 & 1 & 0 & 13 & -3 & & & (1) & \\
\hline
1 & 2 & -1 & 3 & 4 & 0 & & & & \\
0 & 2 & 1 & 4 & -1 & 3 & & & & \\
0 & 0 & 1 & -6 & 5 & -5 & & & & \\
0 & 0 & 0 & -22 & -2 & -2 & 34 & 6 & & \\
0 & 0 & 0 & 34 & -34 & 32 & (22) & & & \\
0 & 0 & 0 & 6 & 8 & 2 & & (22) & & \\
\hline
1 & 2 & -1 & 3 & 4 & 0 & & & & \\
0 & 2 & 1 & 4 & -1 & 3 & & & & \\
0 & 0 & 1 & -6 & 5 & -5 & & & & \\
0 & 0 & 0 & -22 & -2 & -2 & & & & \\
0 & 0 & 0 & 0 & -816 & 636 & 164 & & & \\
0 & 0 & 0 & 0 & 164 & 32 & (816) & & & \\
\end{array}$$

which lead to

$$C = \begin{pmatrix} 1 & 2 & -1 & 3 & 4 & 0 \\ 0 & 2 & 1 & 4 & -1 & 3 \\ 0 & 0 & 1 & -6 & 5 & -5 \\ 0 & 0 & 0 & -22 & -2 & -2 \\ 0 & 0 & 0 & 0 & -816 & 636 \\ 0 & 0 & 0 & 0 & 0 & 130416 \end{pmatrix}$$

Hence

$$\det \mathbf{A} = \frac{(-1) \times 1 \times 2 \times 1 \times -22 \times -816 \times 130416}{1 \times 1 \times 1 \times 2 \times 2 \times 2 \times 2 \times 1 \times 1 \times 1 \times 22 \times 22 \times 816}$$

$$= -741$$

The (-1) pre-multiplying is due to the initial interchange of the first and second rows of the matrix **A**. The reductions took only five stages. Applying the traditional method in [1] required 75 multiplications involving minor determinants and 55 additions or subtractions.

Example 7

Let **A** = $\begin{pmatrix} 3 & 2-i & 4+i \\ 2-i & 6 & i \\ 4+i & i & 3 \end{pmatrix}$

This is a 3 x 3 matrix with complex entries.[Readers not used to complex numbers should note that $i^2 = -1$ and $(a+ib)(a-ib) = a^2 + b^2$]. To find det **A** we proceed with the following reductions:

3	2−i	4+i	2−i	4+i
2−i	6	i	(3)	
4+i	i	3		(3)
3	2−i	4+i		
0	15+4i	5i−9	5i−9	
0	5i−9	−6−8i	(15+4i)	

Hence

C = $\begin{pmatrix} 3 & 2-i & 4+i \\ 0 & 15+4i & 5i-9 \\ 0 & 0 & -114-54i \end{pmatrix}$

det **A** = $\dfrac{3\times(15+4i)(-114-54i)}{3^2\times(15+4i)} = -38-18i$

Example 8

Solve the equation det **A** = 0, where

$$\mathbf{A} = \begin{pmatrix} 1+x & 2 & 3 & 4 \\ 1 & 2+x & 3 & 4 \\ 1 & 2 & 3+x & 4 \\ 1 & 2 & 3 & 4+x \end{pmatrix}$$

(This example was solved in [3], p.504, using the lengthy and tasking traditional method.)

It will be easier to have 1, as a lower multiplier than (1+x). Therefore we interchange the first and last rows and have the following reductions:

1	2	3	4+x	1	1	1+x	
1	2+x	3	4	(1)			
1	2	3+x	4		(1)		
1+x	2	3	4			(1)	
1	2	3	4+x				
0	x	0	-x	2			
0	0	x	-x				
0	-2x	-3x	$-5x-x^2$	(1)			
1	2	3	4+x				
0	x	0	-x				
0	0	x	-x	3			
0	0	-3x	$-7x-x^2$	(1)			

20

Hence

$$C = \begin{pmatrix} 1 & 2 & 3 & 4+x \\ 0 & x & 0 & -x \\ 0 & 0 & x & -x \\ 0 & 0 & 0 & -10x - x^2 \end{pmatrix}$$

$$\det \mathbf{A} = \frac{-x(10+x)x^2(-1)}{1 \times 1 \times 1 \times 1 \times 1}$$

The (-1) appearing is due to the initial interchange of the first and last rows:

Solving det **A** = 0, we have x = 0, 0, 0, -10.

4. Conclusion

The procedure of obtaining the determinant of an n x n matrix by reduction to echelon form as described in this paper is systematic and simple. It has obvious advantage over the traditional method of cofactors which is tedious to apply as n →∞. The beauty of this procedure is easily appreciated in those cases where the square matrices are of high orders greater than three.

5. EXERCISES

The reader is encouraged to attempt the following exercises using Jibunoh's method.

(I) Find det $\mathbf{A} = \begin{vmatrix} 3 & -2 & -5 & 4 \\ -5 & 2 & 8 & -5 \\ -2 & 4 & 7 & -3 \\ 2 & -3 & -5 & 8 \end{vmatrix}$

as obtained from [2]; (Ans = -54).

Confirm with this example, the theorem that, for any square matrix A, det $\mathbf{A}$ = det$\mathbf{A}^T$

(II) Evaluate the determinant of **A**, where

$$A = \begin{pmatrix} \frac{1}{2} & -1 & -\frac{1}{3} \\ \frac{3}{4} & \frac{1}{2} & -1 \\ 1 & -4 & 1 \end{pmatrix} \text{ as obtained from [2].}$$

Hint: 'If every element in any column (or row) of a matrix is multiplied by the same factor, the whole determinant is multiplied by that factor.' Use this to show that

$$24 \det A = \begin{vmatrix} 3 & -6 & -2 \\ 3 & 2 & -4 \\ 1 & -4 & 1 \end{vmatrix}$$

and hence evaluate the determinant by Jibunoh's method: (Ans = $7/6$)

The determinant of the given matrix can also be found by direct application of Jibunoh's method, without using the Hint.

Questions (III) to (V) are obtained from [3];

(III) Evaluate the determinant $\begin{vmatrix} 0 & 1 & 1 & 1 \\ 1 & 0 & 1 & 1 \\ 1 & 1 & 0 & 1 \\ 1 & 1 & 1 & 0 \end{vmatrix}$

(Ans = -3)

(IV) Prove that x = 1 is a root of the equation:

$$\begin{vmatrix} x+2 & 3 & 3 \\ 3 & x+4 & 5 \\ 3 & 5 & x+4 \end{vmatrix} = 0$$

and find the other two roots;

(Ans = 0, -11)

(V) Evaluate

$$\begin{vmatrix} 1+x_1 & x_2 & x_3 & x_4 \\ x_1 & 1+x_2 & x_3 & x_4 \\ x_1 & x_2 & 1+x_3 & x_4 \\ x_1 & x_2 & x_3 & 1+x_4 \end{vmatrix}$$

(Ans $= 1 + x_1 + x_2 + x_3 + x_4$)

(VI) Show that the determinant

$$\begin{vmatrix} a & 0 & c & 0 \\ 0 & a & 0 & c \\ b & 0 & d & 0 \\ 0 & b & 0 & d \end{vmatrix} = (ad - bc)^2$$

Hence or otherwise deduce that $\begin{vmatrix} 6 & 0 & 4 & 0 \\ 0 & 6 & 0 & 4 \\ 3 & 0 & 2 & 0 \\ 0 & 3 & 0 & 2 \end{vmatrix} = 0$

(VII) Let

$$A = \begin{pmatrix} 2 & 4 & 1 & 6 & 7 \\ 3 & 6 & 2 & 4 & 2 \\ 4 & -7 & 2 & 5 & -1 \\ -5 & 4 & 1 & 3 & 4 \\ 0 & 2 & 1 & 4 & 6 \end{pmatrix}$$

Find det **A** (Ans = 1505)

References

1. Burden, R.L and Faires, J.D. *Numerical Analysis (Fifth Edition)* PWS Publishing Company Boston MA. (1993).
2. Lipschutz, S. *Theory and Problems of Linear Algebra*. Schaum's Outline Series McGraw-Hill Book Company, New York (1968).
3. Walker, G. (Ed.) *The Tutorial Algebra Vo l II*, University Tutorial Press Ltd London E C 1 (1968)

INDEX

Abstract, 1

Backward substitutions, 5, 6, 7

Chessboard rule, 3

Cofactors, 1

Diagonal elements, 1, 2, 4, 6, 8, 14

Echelon,
    form, 1, 2, 3, 7, 8, 10, 15, 16
    matrix, 6
    reduction to, 1, 2, 3, 10, 14

Elementary row operations, 2, 3, 7, 8

Formula,
    for e_{nn}, 7
    for the determinant of an n x n matrix, 6, 7

Formulations,
    for the determinant of a 2 x 2 matrix, 2, 5
    for the determinant of a 3 x 3 matrix, 2, 3, 4, 5, 6
    for the determinant of a 4 x 4 matrix, 6
    for the determinant of an n x n matrix, 4, 7

General formulations, 4, 5, 6, 7

Interchanges,
    column, 8, 10, 11
    row, 8, 11, 16, 18, 20

Jibunoh's method, iv, 21, 22

LCMs of matrix entries, 8, 9, 10

Lower multipliers, 1, 2, 3, 4, 6, 7, 8, 10, 20

Matrix entries,
    complex, 2, 19
    real, 2

Pivot row, 3, 14, 16

Procedures of reduction to echelon form,
    the practical, 8, 10, 11, 13, 14
    the theoretical, 8, 10, 12, 13, 15, 16

Reduced matrix, 3, 5

Remarks, 8

Square matrix, 1, 4, 8, 11, 16, 21

Zero entry, 8

www.ingramcontent.com/pod-product-compliance
Lightning Source LLC
Chambersburg PA
CBHW081815170526
45167CB00008B/3444